A New System of Alternating Current Motors and Transformers and Other Essays

A New System of Alternating Current Motors and Transformers and Other Essays
by Nikola Tesla

Wilder Publications, LLC.
PO Box 3005
Radford VA 24143-3005

ISBN 10: 1-934451-78-9
ISBN 13: 978-1-934451-78-6

First Edition

10 9 8 7 6 5 4 3 2 1

TABLE OF CONTENTS

A New System of Alternating Current Motors and Transformers

Delivered before the American Institute of Electrical Engineers, May 1888.

I desire to express my thanks to Professor Anthony for the help he has given me in this matter. I would also like to express my thanks to Mr. Pope and Mr. Martin for their aid. The notice was rather short, and I have not been able to treat the subject so extensively as I could have desired, my health not being in the best condition at present. I ask your kind indulgence, and I shall be very much gratified if the little I have done meets your approval.

In the presence of the existing diversity of opinion regarding the relative merits of the alternate and continuous current systems, great importance is attached to the question whether alternate currents can be successfully utilized in the operation of motors. The transformers, with their numerous advantages, have afforded us a relatively perfect system of distribution, and although, as in all branches of the art, many improvements are desirable, comparatively little remains to be done in this direction. The transmission of power, on the contrary, has been almost entirely confined to the use of continuous currents, and notwithstanding that many efforts have been made to utilize alternate currents for this purpose, they have, up to the present, at least as far as known, failed to give the result desired. Of the various motors adapted to be used on alternate current circuits the following have been mentioned: 1. A series motor with subdivided field. 2. An alternate current generator having its field excited by continuous currents. 3. Elihu Thomson's motor. 4. A combined alternate and continuous current motor. Two more motors of this kind have suggested themselves to me.

1. A motor with one of its circuits in series with a transformer and the other in the secondary of the transformer. 2. A motor having its armature circuit connected to the generator and the field coils closed upon themselves. These, however, I mention only incidentally.

The subject which I now have the pleasure of bringing to your notice is a novel system of electric distribution and transmission of power by means of alternate currents, affording peculiar advantages, particularly in the way of motors, which I am confident will at once establish the superior adaptability of these currents to the transmission of power and will show that many results heretofore unattainable can be reached by their use; results which are very much desired in the practical operation of such systems and which cannot be accomplished by means of continuous currents.

Before going into a detailed description of this system, I think it necessary to make a few remarks with reference to certain conditions existing in continuous current generators and motors, which, although generally known, are frequently disregarded.

In our dynamo machines, it is well known, we generate alternate currents which we direct by means of a commutator, a complicated device and, it may be justly said, the source of most of the troubles experienced in the operation of the machines. Now, the currents so directed cannot be utilized in the motor, but they must—again by means of a similar unreliable device—be reconverted into their original state of alternate currents. The function of the commutator is entirely external, and in no way dues it affect the internal working of the machines. In reality, therefore, all machines are alternate current machines, the currents appearing as continuous only in the external circuit during their transit from generator to motor. In view simply of this fact, alternate currents would commend themselves as a more direct application of electrical energy, and the employment of continuous currents would only be justified if we had dynamos which would primarily generate, and motors which would be directly actuated by such currents.

But the operation of the commutator on a motor is twofold; firstly, it reverses the currents through the motor, and secondly, it effects, automatically, a progressive shifting of the poles of one of its magnetic constituents. Assuming, therefore, that both of the useless operations in the system, that is to say, the directing of the alternate currents on the generator and reversing the direct currents on the motor, be eliminated, it would still be necessary, in order to cause a rotation of the motor, to produce a progressive shifting of the poles of one of its elements, and the question presented itself,—How to perform this operation by the direct action of alternate currents? I will now proceed to show how this result was accomplished.

figures 1, 1a

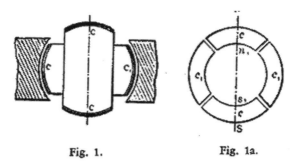

Fig. 1.   Fig. 1a.

In the first experiment a drum-armature was provided with two coils at right angles to each other, and the ends of these coils were connected to two pairs of insulated contact-rings as usual. A ring was then made of thin insulated plates of sheet-iron and wound with four coils, each two opposite coils being connected together so as to produce free poles on diametrically opposite sides of the ring. The remaining free ends of the coils were then connected to the contact-rings of the generator armature so as to form two independent circuits, as indicated in figure 9. It may now be seen what results were secured in this combination, and with this view I would refer to the diagrams, figures 1 to 8a. The field of the generator being independently excited, the rotation of the armature sets up currents in the coils C C1, varying in strength and direction in the well-known manner. In the position shown in figure 1 the current in coil C is nil while coil C1 is traversed by its maximum current, and the connections my be such that the ring is magnetized by the coils c1 c1 as indicated by the letters N S in figure 1a, the magnetizing effect of the coils c c being nil, since these coils are included in the circuit of coil C.

figures 2, 2a

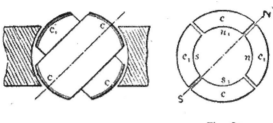

Fig. 2.   Fig. 2a.

In figure 2 the armature coils are shown in a more advanced position, one-eighth of one revolution being completed. Figure 2a illustrates the corresponding magnetic condition of the ring. At this moment the coil c1 generates a current of the same direction as previously, but weaker, producing the poles n1 s1 upon the ring; the coil c also generates a current of the same direction, and the connections may be such that the coils c c produce the poles n s, as shown in figure 2a. The resulting polarity is indicated by the letters N S, and it will be observed that the poles of the ring have been shifted one-eighth of the periphery of the same.

figures 3, 3a

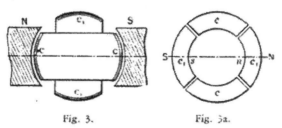

Fig. 3.   Fig. 3a.

In figure 3 the armature has completed one-quarter of one revolution. In this phase the current in coil C is maximum, and of such direction as to produce the poles N S in figure 3a, whereas the current in coil C1 is nil, this coil being at its neutral position. The poles N S in figure 3a are thus shifted one-quarter of the circumference of the ring.

figures 4, 4a

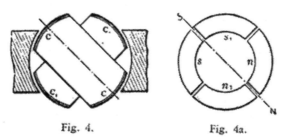

Fig. 4.   Fig. 4a.

Figure 4 shows the coils C C in a still more advanced position, the armature having completed three-eighths of one revolution. At that moment the coil C still generates a current of the same direction as before, but of less strength, producing the comparatively weaker poles n s in figure 4a, The current in the coil C1 is of the same strength, but of opposite direction. Its effect is, therefore, to produce upon the ring the poles n1 and sl as indicated, and a polarity, N S, results, the poles now being shifted three-eighths of the periphery of the ring.

figures 5, 5a

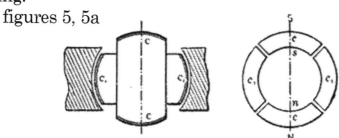

Fig. 5.   Fig. 5a.

In figure 5 one-half of one revolution of the armature is completed, and the resulting magnetic condition of the ring is indicated in figure 5a. Now, the current in coil C is nil, while the coil C1 yields its maximum current, which is of the same direction as previously; the magnetizing effect is, therefore, due to the coils C1 C1 alone, and, referring to figure 5a, it will be observed that the poles N S are shifted one-half of the circumference of the ring. During the next half revolution the operations are repeated, as represented in the figures 6 to 8a.

figures 6, 6a

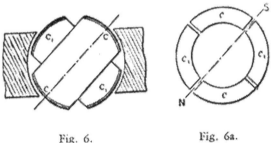

Fig. 6.   Fig. 6a.

A reference to the diagrams will make it clear that during one revolution of the armature the poles of the ring are shifted once around its periphery, and each revolution producing like effects, a rapid whirling of the poles in harmony with the rotation of the armature is the result. If the connections of either one of the circuits in the ring are reversed, the shifting of the poles is made to progress in the opposite direction, but the operation is identically the same. Instead of using four wires, with like result, three wires may be used, one forming a common return for both circuits.

figures 7, 7a

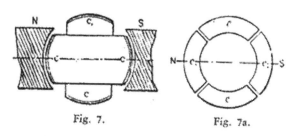

Fig. 7.　　　　　　Fig. 7a.

This rotation or whirling of the poles manifests itself in a series of curious phenomena. If a delicately pivoted disc of steel or other magnetic metal is approached to the ring it is set in rapid rotation, the direction of rotation varying with the position of the disc. For instance, noting the direction outside of the ring it will be found that inside the ring it turns in an opposite direction, while it is unaffected if placed in a position symmetrical to the ring. This is easily explained. Each time that a pole approaches it induces an opposite pole in the nearest point on the disc, and an attraction is produced upon that point; owing to this, as the pole is shifted further away from the disc a tangential pull is exerted upon the same, and the action being constantly repeated, a more or less rapid rotation of the disc is the result. As the pull is exerted mainly upon that part which is nearest to the ring, the rotation outside and inside, or right and left, respectively, is in opposite directions, figure 9. When placed symmetrically to the ring, the pull on opposite sides of the disc being equal, no rotation results. The action is based on the magnetic inertia of the iron; for this reason a disc of hard steel is much more affected than a disc of soft iron, the latter being capable of very rapid variations of magnetism. Such a disc has proved to be a very useful instrument in all these investigations, as it has enabled me to detect any irregularity in the action. A curious effect is also produced upon iron filings. By placing some upon a paper and holding them externally quite close to the ring they are set in a vibrating motion, remaining in the same place, although the paper may be moved back and forth; but in lifting the paper to a certain height which seems to be dependent on the intensity of the poles and the speed of rotation, they are thrown away in a direction always opposite to the supposed movement of the poles. If a paper with filings is put flat upon the ring and the current turned on suddenly; the existence of a magnetic whirl may be easily observed.

figures 8, 8a

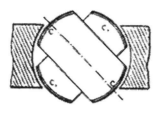
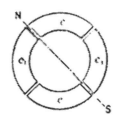

Fig. 8.   Fig. 8a.

To demonstrate the complete analogy between the ring and a revolving magnet, a strongly energized electro-magnet was rotated by mechanical power, and phenomena identical in every particular to those mentioned above were observed.

Obviously, the rotation of the poles produces corresponding inductive effects and may be utilized to generate currents in a closed conductor placed within the influence of the poles. For this purpose it is convenient to wind a ring with two sets of superimposed coils forming respectively the primary and secondary circuits, as shown in figure 10. In order to secure the most economical results the magnetic circuit should be completely closed, and with this object in view the construction may be modified at will.

The inductive effect exerted upon the secondary coils will be mainly due to the shifting or movement of the magnetic action; but there may also be currents set up in the circuits in consequence of the variations in the intensity of the poles. However, by properly designing the generator and determining the magnetizing effect of the primary coils the latter element may be made to disappear. The intensity of the poles being maintained constant, the action of the apparatus will be perfect, and the same result will be secured as though the shifting were effected by means of a commutator with an infinite number of bars. In such case the theoretical relation between the energizing effect of each set of primary coils and their resultant magnetizing effect may be expressed by the equation of a circle having its center coinciding with that of an orthogonal system of axes, and in which the radius represents the resultant and the co-ordinates both of its components. These are then respectively the sine and cosine of the angle U between the radius and one of the axes (O X). Referring to figure 11, we have $r^2 = x^2 + y^2$; where $x = r \cos a$, and $y = r \sin a$.

Assuming the magnetizing effect of each set of coils in the transformer to be proportional to the current—which may be admitted for weak degrees of magnetization—then $x = Kc$ and $y = Kc_1$, where K is a constant and c and c_1 the current in both sets of coils respectively. Supposing, further, the field of the generator to be uniform, we have for constant speed $c_1 = K_1 \sin a$ and $c = K_1 \sin (90o + a) = K_1 \cos a$, where K_1 is a constant. See figure 12. Therefore, $x = Kc = K K_1 \cos a$; $y = Kc_1 = K K_1 \sin a$; and $K K_1 = r$.

That is, for a uniform field the disposition of the two coils at right angles will secure the theoretical result, and the intensity of the shifting poles will be constant. But from $r^2 = x^2 + y^2$ it follows that for $y = O$, $r = x$; it follows that the joint magnetizing effect of both sets of coils should be equal to the effect of one set when at its maximum action. In transformers and in a certain class of motors the fluctuation of the poles is not of great importance, but in another class of these motors it is desirable to obtain the theoretical result.

In applying this principle to the construction of motors, two typical forms of motor have been developed. First, a form having a comparatively small rotary effort at the start, but maintaining a perfectly uniform speed at all loads, which motor has been termed synchronous. Second, a form possessing a great rotary effort at the start, the speed being dependent on the load.

These motors may be operated in three different ways: 1. By the alternate currents of the source only. 2.

By a combined action of these and of induced currents. 3. By the joint action of alternate and continuous currents.

figures 9,

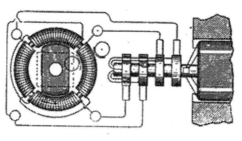

Fig. 9.

The simplest form of a synchronous motor is obtained by winding a laminated ring provided with pole projections with four coils, and connecting the same in the manner before indicated. An iron disc having a segment cut away on each side may be used as an armature. Such a motor is shown in figure 9. The disc being arranged to rotate freely within the ring in close proximity to the projections, it is evident that as the poles are shifted it will, owing to its tendency to place itself in such a position as to embrace the greatest number of the lines of force, closely follow the movement of the poles, and its motion will be synchronous with that of the armature of the generator; that is, in the peculiar disposition shown in figure 9, in which the armature produces by one revolution two current impulses in each of the circuits. It is evident that if, by one revolution of the armature, a greater number of impulses is produced, the speed of the motor will be correspondingly increased. Considering that the attraction exerted upon the disc is greatest when the same is in close proximity to the poles, it follows that such a motor will maintain exactly the same speed at all loads within the limits of its capacity.

Figure 10,

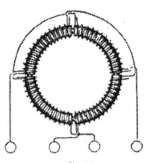

Fig. 10.

To facilitate the starting, the disc may be provided with a coil closed upon itself. The advantage secured by such a coil is evident. On the start the currents set up in the coil strongly energize the disc and increase the attraction exerted upon the same by the ring, and currents being generated in the coil as long as the speed of the armature is inferior to that of the poles, considerable work may be performed by such a motor even if the speed be below normal. The intensity of the poles being constant, no currents will be generated in the coil when the motor is turning at its normal speed.

figures 11. 12

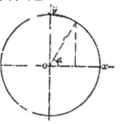
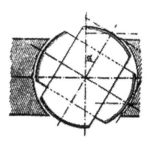

Fig. 11.   Fig. 12.

Instead of closing the coil upon itself, its ends may be connected to two insulated sliding rings, and a continuous current supplied to these from a suitable generator. The proper way to start such a motor is to close the coil upon itself until the normal speed is reached, or nearly so, and then turn on the continuous current. If the disc be very strongly energized by a continuous current the motor may not be able to start, but if it be weakly energized, or generally so that the magnetizing effect of the ring is preponderating it will start and reach the normal speed. Such a motor will maintain absolutely the same speed at all loads. It has also been found that if the motive power of the generator is not excessive, by checking the motor the speed of the generator is diminished in synchronism with that of the motor. It is characteristic of this form of motor that it cannot be reversed by reversing the continuous current through the coil.

figure 13

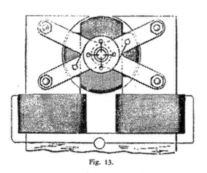

Fig. 13.

The synchronism of these motors may be demonstrated experimentally in a variety of ways. For this purpose it is best to employ a motor consisting of a stationary field magnet and an armature arranged to rotate within the same, as indicated in figure 13. In this case the shifting of the poles of the armature produces a rotation of the latter in the opposite direction. It results therefrom that when the normal speed is reached, the poles of the armature assume fixed positions relatively to the field magnet and the same is magnetized by induction, exhibiting a distinct pole on each of the pole-pieces. If a piece of soft iron is approached to the field magnet it will at the start be attracted with a rapid vibrating motion produced by the reversals of polarity of the magnet, but as the speed of the armature increases; the vibrations become less and less frequent and finally entirely cease. Then the iron is weakly but permanently attracted, showing that the synchronism is reached and the field magnet energized by induction.

The disc may also be used for the experiment. If held quite close to the armature it will turn as long as the speed of rotation of the poles exceeds that of the armature; but when the normal speed is reached, or very nearly so; it ceases to rotate and is permanently attracted.

A crude but illustrative experiment is made with an incandescent lamp. Placing the lamp in circuit with the continuous current generator, and in series with the magnet coil, rapid fluctuations are observed in the light in consequence of the induced current set up in the coil at the start; the speed increasing, the fluctuations occur at longer intervals, until they entirely disappear, showing that the motor has attained its normal speed. A telephone receiver affords a most sensitive instrument; when connected to any circuit in the motor the synchronism may be easily detected on the disappearance of the induced currents.

In motors of the synchronous type it is desirable to maintain the quantity of the shifting magnetism constant, especially if the magnets are not properly subdivided.

To obtain a rotary effort in these motors was the subject of long thought. In order to secure this result it was necessary to make such a disposition that while the poles of one element of the motor are shifted by the alternate currents of the source, the poles produced upon the other element should always be maintained in the proper relation to the former, irrespective of the speed of the motor. Such a condition exists in a continuous current motor; but in a synchronous motor, such as described, this condition is fulfilled only when the speed is normal.

figure 14

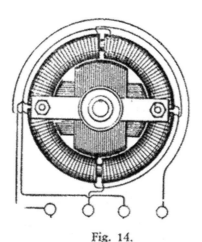

Fig. 14.

The object has been attained by placing within the ring a properly subdivided cylindrical iron core wound with several independent coils closed upon themselves.

Two coils at right angles as in figure 14, are sufficient, but greater number may he advantageously employed. It results from this disposition that when the poles of the ring are shifted, currents are generated in the closed armature coils. These currents are the most intense at or near the points of the greatest density of the lines of force, and their effect is to produce poles upon the armature at right angles to those of the ring, at least theoretically so; and since action is entirely independent of the speed—that is, as far as the location of the poles is concerned—a continuous pull is exerted upon the periphery of the armature. In many respects these motors are similar to the continuous current motors. If load is put on, the speed, and also the resistance of the motor, is diminished and more current is made to pass through the energizing coils, thus increasing the effort. Upon the load being taken off, the counter-electromotive force increases and less current passes through the primary or energizing coils. Without any load the speed is very nearly equal to that of the shifting poles of the field magnet.

It will be found that the rotary effort in these motors fully equals that of the continuous current motors. The effort seems to be greatest when both armature and field magnet are without any projections; but as in such dispositions the field cannot be very concentrated, probably the best results will be obtained by leaving pole projections on one of the elements only. Generally, it may be stated that the projections diminish the torque and produce a tendency to synchronism.

A characteristic feature of motors of this kind is their capacity of being very rapidly reversed. This follows from the peculiar action of the motor. Suppose the armature to be rotating and the direction of rotation of the poles to be reversed. The apparatus then represents a dynamo machine, the power to drive this machine being the momentum stored up in the armature and its speed being the sum of the speeds of the armature and the poles.

If we now consider that the power to drive such a dynamo would be very nearly proportional to the third power of the speed, for this reason alone the armature should be quickly reversed. But simultaneously with the reversal another element is brought into action, namely, as the movement of the poles with respect to the armature is reversed, the motor acts like a transformer in which the resistance of the secondary circuit would be abnormally diminished by producing in this circuit an additional electromotive force. Owing to these causes the reversal is instantaneous.

If it is desirable to secure a constant speed, and at the same time a certain effort at the start, this result may be easily attained in a variety of ways. For instance, two armatures, one for torque and the other for synchronism, may be fastened on the same shaft, and any desired preponderance may be given to either one, or an armature may be wound for rotary effort, but a more or less pronounced tendency to synchronism may be given to it by properly constructing the iron core; and in many other ways.

As a means of obtaining the required phase of the currents in both the circuits, the disposition of the two coils at right angles is the simplest, securing the most uniform action; but the phase may be obtained in many other ways, varying with the machine employed. Any of the dynamos at present in use may be easily adapted for this purpose by making connections to proper points of the generating coils. In closed circuit armatures, such as used in the continuous current systems, it is best to make four derivations from equi-distant points or bars of the commutator, and to connect the same to four insulated sliding rings on the shaft. In this case each of the motor circuits is connected to two diametrically opposite bars of the commutator. In such a disposition the motor may also be operated at half the potential and on the three-wire plan, by connecting the motor circuits in the proper order to three of the contact rings.

In multipolar dynamo machines, such as used in the converter systems, the phase is conveniently obtained by winding upon the armature two series of coils in such a manner that while the coils of one set or series are at their maximum production of current, the coils of the other will be at their neutral position, or nearly so, whereby both sets of coils may be subjected simultaneously or successively to the inducing action of the field magnets.

figures 15, 16, 17

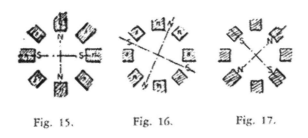

Fig. 15.   Fig. 16.   Fig. 17.

Generally the circuits in the motor will be similarly disposed, and various arrangements may be made to fulfill the requirements; but the simplest and most practicable is to arrange primary circuits on stationary parts of the motor, thereby obviating, at least in certain forms, the employment of sliding contacts. In such a case the magnet coils are connected alternately in both the circuits; that is 1, 3, 5 . . . in one, and 2, 4, 6 . . . in the other, and the coils of each set of series may be connected all in the same manner, or alternately in opposition; in the latter case a motor with half the number of poles will result, and its action will be correspondingly modified. The figures 15, 16 and 17, show three different phases, the magnet coils in each circuit being connected alternately in opposition. In this case there will be always four poles, as in figures 15 and 17, four pole projections will be neutral, and in figure 16 two adjacent pole projections will have the same polarity. If the coils are connected in the same manner there will be eight alternating poles, as indicated by the letters n' s' in fig.15.

The employment of multipolar motors secures in this system an advantage much desired and unattainable in the continuous current system, and that is, that a motor may be made to run exactly at a predetermined speed irrespective of imperfections in construction, of the load, and, within certain limits, of electromotive force and current strength.

In a general distribution system of this kind the following plan should be adopted. At the central station of supply a generator should be provided having a considerable number of poles. The motors operated from this generator should be of the synchronous type, but possessing sufficient rotary effort to insure their starting. With the observance of proper rules of construction it may be admitted that the speed of each motor will be in some inverse proportion to its size, and the number of poles should be chosen accordingly. Still exceptional demands may modify this rule. In view of this, it will be advantageous to provide each motor with a greater number of pole projections or coils, the number being preferably a multiple of two and three. By this means, by simply changing the connections of the coils, the motor may be adapted to any probable demands.

If the number of the poles in the motor is even, the action will he harmonious and the proper result will be obtained; if this is not the case the best plan to be followed is to make a motor with a double number of poles and connect the same in the manner before indicated, so that half the number of poles result. Suppose, for instance, that the generator has twelve poles, and it would be desired to obtain a speed equal to 12/7 of the speed of the generator. This would require a motor with seven pole projections or magnets, and such a motor could not be properly connected in the circuits unless fourteen armature coils would be provided, which would necessitate the employment of sliding contacts. To avoid this the motor should be provided with fourteen magnets and seven connected in each circuit, the magnets in each circuit alternating among themselves. The armature should have fourteen closed coils. The action of the motor will not be quite as perfect as in the case of an even number of poles, but the drawback will not be of a serious nature.

However, the disadvantages resulting from this unsymmetrical form will be reduced in the same proportion as the number of the poles is augmented.

If the generator has, say, n, and the motor n1 poles, the speed of the motor will be equal to that of the generator multiplied by n/n1.

figures 18, 19, 20, 21

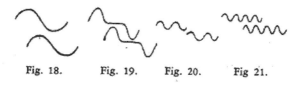

Fig. 18.   Fig. 19.   Fig. 20.   Fig 21.

The speed of the motor will generally be dependent on the number of the poles, but there may be exceptions to this rule. The speed may be modified by the phase of the currents in the circuits or by the character of the current impulses or by intervals between each or between groups of impulses. Some of the possible cases are indicated in the diagrams, figures 18, 19, 20 and 21, which are self-explanatory. Figure 18 represents the condition generally existing, and which secures the best result. In such a case, if the typical form of motor illustrated in figure 9 is employed, one complete wave in each circuit will produce one revolution of the motor. In figure 19 the same result will he effected by one wave in each circuit, the impulses being successive; in figure 20 by four, and in figure 21 by eight waves.

By such means any desired speed may be attained; that is, at least within the limits of practical demands. This system possesses this advantage besides others, resulting from simplicity. At full loads the motors show efficiency fully equal to that of the continuous current motors. The transformers present an additional advantage in their capability of operating motors. They are capable of similar modifications in construction, and will facilitate the introduction of motors and their adaptation to practical demands. Their efficiency should be higher than that of the present transformers, and I base my assertion on the following:

In a transformer as constructed at present we produce the currents in the secondary circuit by varying the strength of the primary or exciting currents. If we admit proportionality with respect to the iron core the inductive effect exerted upon the secondary coil will be proportional to the numerical sum of the variations in the strength of the exciting current per unit of time; whence it follows that for a given variation any prolongation of the primary current will result in a proportional loss. In order to obtain rapid variations in the strength of the current, essential to efficient induction, a great number of undulations are employed. From this practice various disadvantages result. These are, increased cost and diminished efficiency of the generator, more waste of energy in heating the cores, and also diminished output of the transformer, since the core is not properly utilized, the reversals being too rapid. The inductive effect is also very small in certain phases, as will be apparent from a graphic representation, and there may be periods of inaction, if there are intervals between the succeeding current impulses or waves. In producing a shifting of the poles in the transformer, and thereby inducing currents, the induction is of the ideal character, being always maintained at its maximum action. It is also reasonable to assume that by a shifting of the poles less energy will be wasted than by reversals.

MR. NIKOLA TESLA ON ALTERNATING CURRENT MOTORS

Electrical World — N. Y. — May 25, 1889

To the Editor of The Electrical World:

Sir: About a year ago I had the pleasure of bringing before the American Institute of Electrical Engineers the results of some of my work on alternate current motors. They were received with the interest which novel ideas never fail to excite in scientific circles, and elicited considerable comment. With truly American generosity, for which, on my part, I am ever thankful, a great deal of praise through the columns of your esteemed paper and other journals has been bestowed upon the originator of the idea, in itself insignificant. At that time it was impossible for me to bring before the Institute other results in the same line of thought. Moreover, I did not think it probable — considering the novelty of the idea — that anybody else would be likely to pursue work in the same direction. By one of the most curious coincidences, however, Professor Ferraris not only came independently to the same theoretical results, but in a manner identical almost to the smallest detail. Far from being disappointed at being prevented from calling the discovery of the principle exclusively my own, I have been excessively pleased to see my views, which I had formed and carried out long before, confirmed by this eminent man, to whom I consider myself happy to be related in spirit, and toward whom, ever since the knowledge of the facts has reached me, I have entertained feelings of the most sincere sympathy and esteem. In his able essay Prof. Ferraris omitted to mention various other ways of accomplishing similar results, some of which have later been indicated by O. B. Shallenberger, who some time before the publication of the results obtained by Prof. Ferraris and myself had utilized the principle in the construction of his now well known alternate current meter, and at a still later period by Prof. Elihu Thomson and Mr. M. J. Wightman.

Since the original publications, for obvious reasons, little has been made known in regard to the further progress of the invention; nevertheless the work of perfecting has been carried on indefatigably with all the intelligent help and means which a corporation almost unlimited in its resources could command, and marked progress has been made in every direction. It is therefore not surprising that many unaquainted with this fact, in expressing their views as to the results obtained, have grossly erred.

In your issue of May 4, I find a communication from the electricians of Ganz & Co., of Budapest, relating to certain results observed in recent experiments with a novel form of alternate current motor. I would have nothing to say in regard to this communication unless it were to sincerely congratulate these gentlemen on any good results which they may have obtained, but for the article, seemingly inspired by them, which appeared in the London Electrical Review of April 26, wherein certain erroneous views are indorsed and some radically false assertions made, which, though they may be quite unintentional, are such as to create prejudice and affect material interests.

As to the results presented, they not only do not show anything extraordinary, but are, in fact, considerably below some figures obtained with my motors a long time ago. The main stress being laid upon the proposition between the apparent and real energy supplied, or perhaps more directly, upon the ratio of the energy apparently supplied to, and the real energy developed by, the motor, I will here submit, with your permission, to your readers, the results respectively arrived at by these gentlemen and myself.

Energy apparently supplied in watts.		Work performed in watts.		Ratio of energy apparently supplied to the real energy developed.	
Ganz & Co.	Westinghouse Co.	Ganz & Co.	Westinghouse Co.	Ganz & Co.	Westinghouse Co.
18,000	21,840	11,000	17,595	0.611	0.805
24,200	30,295	14,600	25,365	0.603	0.836
29,800	43,624	22,700	36,915	0.761	0.816
......	56,800	......	48,675	.....	0.856
......	67,500	......	59,440	.....	0.88
......	79,100	......	67,365	.....	0.851

If we compare these figures we will find that the most favorable ratio in Ganz & Co's motor is 0.761, whereas in the Westinghouse, for about the same load, it is 0.836, while in other instances, as may be seen, it is still more favorable. Notwithstanding this, the conditions of the test were not such as to warrant the best possible results.

The factors upon which the apparent energy is mainly dependent could have been better determined by a proper construction of the motor and observance of certain conditions. In fact, with such a motor a current regulation may be obtained

which, for all practical purposes, is as good as that
of the direct current motors, and the only
disadvantage, if it be one, is that when the motor is
running without load the apparent energy cannot be
reduced quite as low as might be desirable. For
instance, in the case of this motor the smallest
amount of apparent energy was about 3,000 watts,
which is certainly not very much for a machine
capable of developing 90 h. p. of work; besides, the
amount could have been reduced very likely to 2,000
watts or less.

On the other hand, these motors possess the
beautiful feature of maintaining an absolutely
constant speed no matter how the load may vary.
This feature may be illustrated best by the following
experiment performed with this motor. The motor
was run empty, and a load of about 200 h. p.., far
exceeding the normal load, was thrown on suddenly.
Both armatures of the motor and generator were
seen to stop for an instant, the belts slipping over
the pulleys, whereupon both came up to the normal
speed with the full load, not having been thrown out
of synchronism. The experiment could be repeated
any number of times. In some cases, the driving
power being sufficient, I have been enabled to throw
on a load exceeding 8 to 9 times that which the
motor was designed to carry, without affecting the
speed in the least.

This will be easily understood from the manner in
which the current regulation is effected. Assuming
the motor to be running without any load, the poles
of the armature and field have a certain relative
position which is that of the highest self-induction or
counter electromotive force. If load be thrown on,

the poles are made to recede; the self-induction or counter electromotive force is thereby diminished and more current passed trough the stationary or movable armature-coils. This regulation is very different from that of a direct current motor. In the latter the current is varied by the motor losing a certain number of revolutions in proportion to the load, and the regulation would be impossible if the speed would be maintained constant; here the whole regulation is practically effected during a fraction of one revolution only. From this it is also apparent that it is a practical impossibility to throw such a motor out of synchronism, as the whole work must be done in an instant, it being evident that if the load is not sufficient to make a motor lose a fraction of the first revolution it will not be able to do so in the succeeding revolutions. As to the efficiency of these motors, it is perfectly practicable to obtain 94 to 95 per cent.

The results above given were obtained on a three-wire system. The same motor has been started and operated on two wires in a variety of ways, and although it was not capable of performing quite as much work as on three wires, up to about 60 h. p. it gave results practically the same as those above-mentioned. In fairness to the electricians of Ganz & Co., I must state here that the speed of this motor was higher than that used in their experiments, it being about 1,500. I cannot make due allowance for this difference, as the diameter of the armature and other particulars of the Ganz & Co. motor were not given.

The motor tested had a weight of about 5,000 lbs. From this it will be seen that the performance even on two wires was quite equal to that of the best direct current motors. The motor being of a synchronous type, it might be implied that it was not capable of starting. On the contrary, however, it had a considerable torque on the start and was capable of starting under fair load.

In the article above referred to the assertion is made that the weight of such alternate current motor, for a given capacity, is "several times" larger than that of a direct current motor. In answer to this I will state here that we have motors which with a weight of about 850 pounds develop 10 h. p. with an efficiency of very nearly 90 per cent, and the spectacle of a direct current motor weighing, say 200 – 300 pounds and performing the same work, would be very gratifying for me to behold. The motor which I have just mentioned had no commutator or brushes of any kind nor did it require any direct current.

Finally, in order to refute various assertions made at random, principally in the foreign papers, I will take the liberty of calling to the attention of the critics the fact that since the discovery of the principle several types of motors have been perfected and of entirely different characteristics, each suited for a special kind of work, so that while one may be preferable on account of its ideal simplicity, another might be more efficient. It is evidently impossible to unite all imaginable advantages in one form, and it is equally unfair and unreasonable to judge all different forms according to a common standard. Which form of the existing motors is best, time will show; but even in the

present state of the art we are enabled to satisfy any
possible demand in practice.
  Nikola Tesla
  Pittsburgh, Pa.

Tesla's New Alternating Motors

The Electrical Engineer - N. Y. — Sept. 24, 1890

I hope you will allow me the privilege, to say in the columns of your esteemed journal a few words in regard to an article which appeared in *Industries* of August 22, to which my attention has been called. In this article an attempt is made to criticise some of my inventions, notably those which you have described in your issue of August 6, 1890.

The writer begins by stating: "The motor depends on a shifting of the poles under certain conditions, a principle which has been *already* employed by Mr. A. Wright in his alternating current meter." This is no surprise to me. It would rather have surprised me to learn that Mr. Wright has *not yet* employed the principle in his meter, considering what, before its appearance, was known of my work on motors, and more particularly of that of Schallenberger on meters. It has cost me years of thought to arrive at certain results, by many believed to be unattainable), for which there are now numerous claimants, and the number of these is rapidly increasing, like that of the colonels in the South after the war.

The writer then good-naturedly explains the theory of action of the motive device in Wright's meter, which has greatly benefited me, for it is so long since I have arrived at this, and similar theories, that I had almost forgotten it. He then says: " Mr. Tesla has worked out some more or less complicated motors on this principle, but the curious point is that he has completely misunderstood the theory of the phenomena, and has got hold of (the old fallacy of screening." This may be *curious,* but how much *more curious* it is to find that the writer in *Industries* has *completely misunderstood* everything himself. I like nothing better than just criticism of my work, even if it be severe, but when the critic assumes a certain " l'état c'est moi" air of unquestioned Competency I want him to know what he is writing about. How little the writer in *Industries* seems to know about the matter is painfully apparent when ho connects the phenomenon in Wright's meter with the subject he has under consideration. His further remark, "He (Mr. Tesla) winds his secondary of iron instead of copper and thinks the effect is produced magnetically," is illustrative of the care with which he has perused the description of the devices contained in the issue of *The Electrical Engineer* above referred to.

I take a motor having, say eight poles, and wrap the exciting coils of four alternate cores with fine insulated iron wire. When the current is started in these coils it encounters the effect of the closed magnetic circuit and is retarded. The magnetic lines set up at the start close to the iron wire around the coils and no free poles appear at first at the ends of the four cores. As the current rises in the coils more lines are set up, which crowd more and more in the fine iron wire until finally the same becomes saturated, or nearly so, when the shielding action of the iron wire ceases and free poles appear at the ends of the four protected cores. The effect of the iron wire, as will be seen, is two-fold. First, it retards the energizing current; and second, it delays the appearance of the free poles. To produce still greater difference of phase in the magnetization of the protected and unprotected cores, I connect the iron wire surrounding the coils of the former in series with the coils of the latter, in which case, of course, the iron wire is preferably wound or connected differentially, after the fashion of the resistance, coils in a bridge, so as to have no appreciable self-induction. In other cases I obtain the desired retardation in the appearance of the free poles on one set of cores by a magnetic shunt, which produces a greater retardation of the current and takes up at the start a certain number of the lines set up, but becomes saturated when the current in the exciting coils reaches a predetermined strength.

In the transformer the same principle of shielding is utilized. A primary conductor is surrounded with a fine layer of laminated iron, consisting of fine iron wire or plates properly insulated and interrupted. As long as the current in the primary conductor is so small that the iron enclosure can carry all the lines of force set up by the current, there is very little action exerted upon a secondary conductor placed in vicinity to the first; but just as soon as the iron enclosure becomes saturated, or nearly so, it loses the virtue of protecting the secondary and the inducing action of the primary practically begins. What, may I ask, has all this to do with the "old fallacy of screening?"

With certain objects in view — the enumeration of which would lead me too far — an arrangement was shown in *The Electrical Engineer*, about which the writer in *Industries* says : " A ring of laminated iron is wound with a secondary. It is then encased in iron laminated in the *wrong direction* and the primary is wound outside of this. The layer of iron between the primary and secondary is supposed to screen the coil. Of course it cannot do so, such a thing Is unthinkable." This reminds me of the man who had committed some offense and engaged the services of an attorney. "They cannot commit you to prison for that," said the attorney. Finally the man *was* imprisoned. He sent for the attorney. " Sir," said the latter, " I tell you they *cannot* imprison you for that." "But, sir," retorted the prisoner, "they *have* imprisoned me." It *may not* screen, in the opinion of the writer in *Industries,* but just the same it *does.* According to the arrangement the *principal* effect of the screen may be either a retardation of the action of the primary current upon the secondary circuit or a deformation of the secondary current wave with similar results for the purposes intended. In the arrangement referred to by the writer in *Industries* be seems to be certain that the iron layer acts like a choking coil; there again he is mistaken; it does not act like a choking coil, for then its capacity for maintaining constant current would be very limited. But it acts more like a magnetic shunt in constant current transformers and dynamos, as, in my opinion, it ought to act.

There are a good many more things to be said about the remarks contained in *Industries.* In regard to the magnetic time lag the writer says: " If a bar of iron has a coil at one end, and if the core is perfectly laminated, on starting a current in the coil the induction *all along the iron* corresponds to the excitation at that instant, unless there is a microscopic time lag, *of which there is no evidence."* Yet a motor was described, the very operation of which is dependent on the time lag of magnetization of the different parts of a core. It is true the writer uses the term "perfectly laminated" (which, by the way, I would like him to explain), but if he intends to make such a "perfectly laminated" core I venture to say there is trouble in store for him. From his remarks I see that the writer ccompletely overlooks the importance of the size of the core and of the number of the alternations pointed out; be fails to see the stress laid on the saturation of the screen, or shunt, in some of the cases described; he does not seem to recognize the fact that in the cases considered the formation of current is reduced as far as practicable in the screen, and that the same, therefore, so far as its quality of screening is concerned, has no role to perform as *a conductor.* I also see that he would want considerable information about the time lag in the magnetization of the different parts of a core, and an explanation why, in the transformer he refers to, the screen is laminated in the *wrong direction,* etc. — but the elucidation of all these points would require more time than I am able to devote to the subject. It is distressing to find all this in the columns of a leading technical journal.

In conclusion, the writer shows his true colors by making the following withering remarks: "It is questionable whether the Tesla motor will ever be a success. Such motors will go round, of course, and will give outputs, but their efficiency is doubtful; and if they need three-wire circuits and special generators there is no object in using them, as a direct current motor can be run instead with advantage."

No man of broad views will feel certain of the success of any invention, however good and original, in this period of feverish activity, when every day may bring new and unforseen developments. At the pace we are progressing the permanence of all our apparatus it its present forms becomes more and more problematical. It is impossible to foretell what type of motor will crystalize out of the united efforts of many able men; but it is my conviction that at no distant time a motor having commutator and brushes will be looked upon as an antiquated piece of mechanism. Just how much the last quoted remarks of the writer of *Industries* — considering the present state of the art — are justified, I will endeavor to show in a few lines.

First, take the transmission of power in isolated places. A case frequently occurring in practice and attracting more and more the attention of engineers is the transmission of large powers at considerable distances. In such a case the power is very likely to be cheap, and the cardinal requirements are then the reduction of the cost of the leads, cheapness of construction and maintenance of machinery and constant speed of the motors. Suppose a loss of only 25 per cent in the leads, at full load, be allowed. If a direct current motor be used, there will be, besides other difficulties, considerable variation in the speed of the motor — even if the current is supplied from a series dynamo — so much so that the motor may not be well adapted for many purposes, for instance, in cases where direct current transformation is contemplated with the object of running lights or other devices at constant potential. It is true that the condition may be bettered by employing proper regulating devices, but these will only further complicate the already complex system, and in all probability fail to secure such perfection as will be desired. In using an ordinary single-circuit alternate current motor the disadvantage is that the motor has no starting torque and that, for equal weight, its output and efficiency are more or less below that of a direct current motor. If, on the contrary, the armature of any alternator or direct current machine — large, low-speed, two-pole machines will give the best results — is wound with two circuits, a motor is at once obtained which possesses sufficient torque to start under considerable load: it runs in absolute synchronism with the generator — an advantage much desired and hardly ever to be attained with regulating devices; it takes current in proportion to the load, and its plant efficiency within a few per cent is equal to that of a direct current motor of the same size. It will be able, however, to perform more work than a direct current motor of the same size, first, because there will be no change of speed, even if the load be doubled or tripled, within the limits of available generator power; and second, because it can be run at a higher electromotive force, the commutator and the complication and difficulties it involves in the construction and operation of the generators and motors being eliminated from the system. Such a system will, of course, require three leads, but since the plant efficiency is practically equal to that of the direct current system, it will require the same amount of copper which would be required in the latter system, and the disadvantage of the third lead will be comparatively small, if any, for three leads of smaller size may perhaps be more convenient to place than two larger leads. When more machines have to be used there may be no disadvantage whatever connected with the third wire ; however, since the simplicity of the generators and motors allows the use of higher electromotive forces, the cost of the leads may be reduced below the figure practicable with the direct current system.

Considering all the practical advantages offered by such an alternating system, I am of an opinion quite contrary to that of the author of the article in *Industries,* and think that it can quite successfully stand the competition of any direct current system, and this the more, the larger the machines built and the greater the distances.

Another case frequently occurring in practice is the transmission of small powers in numerous isolated places, such as mines, etc. In many of these cases simplicity and reliability of the apparatus are the principal objects. I believe that in many places of this kind my motor has so far proved a perfect success. In such cases a type of motor is used possessing great starting torque, requiring for its operation only alternating current and having no sliding contacts whatever on the armature, this advantage over other types of motors being highly valued in such places. The plant efficiency of this form of motor is, in the present state of perfection, inferior to that of the former form, but I am confident that improvements will be made in that direction. Be-sides, plant efficiency is in these cases of secondary importance, and in cases of transmission at considerable distances, it is no drawback, since the electromotive force may be raised as high as practicable on converters. I can not lay enough stress on this advantageous feature of my motors, and should think that it ought to be fully appreciated by engineers, for to high electromotive forces we are surely coming, and if they must be used, then the fittest apparatus will be employed. I believe that in the transmission of power with such commutatorless machines, 10,000 volts, and even more, may be used, and I would be glad to see Mr. Ferranti's enterprise succeed. His work is in the right direction, and, in my opinion, it will be of great value for the advancement of the art.

As regards the supply of power from large central stations in cities or centres of manufacture, the above arguments are applicable, and I see no reason why the three-wire motor system should not be successful. In putting up such a station, the third wire would be but a very slight drawback, and the system possesses enough advantages to over-balance this and any other disadvantage. But this question will be settled in the future, for as yet comparatively little has been done in that direction, even with the direct current system. The plant efficiency of such a three-wire system would be increased by using, in connection with the ordinary type of my motor, other types which act more like inert resistances. The plant efficiency of the whole system would, in all cases, be greater than that of each individual motor — if like motors are used — owing to the fact that they would possess different self-induction, according to the load.

The supply of power from lighting mains is, I believe, in the opinion of most engineers, limited to comparatively small powers, for obvious reasons. As the present systems are built on the two-wire plan, an efficient two-wire motor without commutator is required for this purpose, and also for traction purposes. A large number of these motors, embodying new principles, have been devised by me and are being constantly perfected. On lighting stations, however, my three-wire system may be advantageously carried out. A third wire may be run for motors and the old connections left undisturbed. The armatures of the generators may be rewound, whereby the output of the machines will be increased about 35 per cent, or even more in machines with cast iron field magnets. If the machines are worked at the same capacity, this means an increased efficiency. If power is available at the station, the gain in current may be used in motors. Those who object to the third wire, may remember that the old two-wire direct system is almost entirely superseded by the three-wire system, yet my three-wire system offers to the alternating system relatively greater advantages, than the three-wires direct possesses over the two-wire. Perhaps, if the writer in *Industries* would have taken all this in consideration, he would have been less hasty in his conclusions.

Nikola Tesla
New York
Sept. 17, 1890